T0155604

Cambridge Elements ☰

Elements in Geochemical Tracers in Earth System Science
edited by
Timothy Lyons
University of California
Alexandra Turchyn
University of Cambridge
Chris Reinhard
Georgia Institute of Technology

LITHIUM ISOTOPES

A Tracer of Past and Present Silicate Weathering

Philip A. E. Pogge von Strandmann
Johannes Gutenberg University and University College London

Mathieu Dellinger
Durham University

A. Joshua West
University of Southern California

CAMBRIDGE
UNIVERSITY PRESS

University Printing House, Cambridge CB2 8BS, United Kingdom

One Liberty Plaza, 20th Floor, New York, NY 10006, USA

477 Williamstown Road, Port Melbourne, VIC 3207, Australia

314–321, 3rd Floor, Plot 3, Splendor Forum, Jasola District Centre, New Delhi – 110025, India

103 Penang Road, #05–06/07, Visioncrest Commercial, Singapore 238467

Cambridge University Press is part of the University of Cambridge.

It furthers the University's mission by disseminating knowledge in the pursuit of education, learning, and research at the highest international levels of excellence.

www.cambridge.org
Information on this title: www.cambridge.org/9781108964968
DOI: 10.1017/9781108990752

© Philip A. E. Pogge von Strandmann, Mathieu Dellinger and A. Joshua West 2021

First published 2021

A catalogue record for this publication is available from the British Library.

ISBN 978-1-108-96496-8 Paperback
ISSN 2515-7027 (online)
ISSN 2515-6454 (print)

Lithium Isotopes

A Tracer of Past and Present Silicate Weathering

Elements in Geochemical Tracers in Earth System Science

DOI: 10.1017/9781108990752
First published online: August 2021

Philip A. E. Pogge von Strandmann
Johannes Gutenberg University and University College London

Mathieu Dellinger
Durham University

A. Joshua West
University of Southern California

Author for correspondence: Philip A. E. Pogge von Strandmann,
p.strandmann@ucl.ac.uk

Abstract: Lithium isotopes are a relatively novel tracer of present and past silicate weathering processes. Given that silicate weathering is the primary long-term method by which CO_2 is removed from the atmosphere, Li isotope research is going through an exciting phase. We show the weathering processes that fractionate dissolved and sedimentary Li isotope ratios, focusing on weathering intensity and clay formation. We then discuss the carbonate and silicate archive potential of past seawater δ^7Li. These archives have been used to examine Li isotope changes across both short and long timescales. The former can demonstrate the rates at which the climate is stabilized from perturbations via weathering, a fundamental piece of the puzzle of the long-term carbon cycle.

Keywords: weathering, erosion, carbon, climate, oceans

ISBNs: 9781108964968 (PB), 9781108990752 (OC)
ISSNs: 2515-7027 (online), 2515-6454 (print)

Contents

Introduction

Chemical weathering of continental rocks is one of the fundamental Earth system processes that affects climate and ocean chemistry. The weathering of silicate rocks removes carbon from the atmosphere, transporting it as alkalinity together with associated cations (most importantly Ca and Mg) to the oceans (Walker et al., 1981; West et al., 2005). The delivery of alkalinity enhances ocean uptake of CO_2 over centennial timescales and drives burial of carbonate and resulting CO_2 sequestration on timescales of thousands of years. At the same time, weathering provides critical nutrients to the coastal oceans, where they promote primary productivity. Burial of the resulting organic carbon is enhanced via association with clay minerals (Kennedy and Wagner, 2011), which are also provided by chemical weathering. In summary, silicate weathering is one of the primary controllers of the Earth's climate on various timescales.

It is not a surprise, then, that a great deal of research seeking to understand the evolution of Earth's climate, as well as the nutrient fluxes that support life, has focused on evaluating the processes that control weathering. Such controls are generally thought to be either climate-related (temperature, runoff, vegetation) or erosion-related (supply of fresh rock to be weathered) (West et al., 2005). The former would allow a temperature-controlled feedback on weathering (i.e. weathering accelerates during warming, removing more CO_2 and cooling the climate, and vice versa), known as the "weathering thermostat" that keeps the long-term climate relatively stable (Walker et al., 1981). The latter would drive long-term cooling processes when mineral supply increases, for example during mountain-building events (Raymo et al., 1988).

For this reason, a lot of effort has gone into characterizing and quantifying weathering processes both in the present and in Earth's geological past. Lithium isotopes have great potential in this field. Lithium is moderately incompatible during igneous processes and highly fluid mobile during surface processes (Penniston-Dorland et al., 2017; Tomascak et al., 2016). It tends to be concentrated in the continental crust relative to the mantle, and even more concentrated in clays and other secondary products of weathering (Teng et al., 2010).

Lithium has two stable isotopes (^{6}Li and ^{7}Li), the ratio of which, as for other stable isotope systems, is reported in the delta notation, as parts per thousand deviation from the L-SVEC standard: δ^7Li. The pioneering work of Lui-Heung Chan, starting in the late 80s, showed that Li isotopes strongly fractionate during low-temperature clay formation. Chan focused on alteration of the oceanic crust and reverse weathering in the oceans, showing that clays preferentially take up the light Li isotope (^{6}Li), driving residual waters (including the oceans) isotopically heavy (Chan et al., 1992). This concept was taken forward into examining river waters by Youngsook Huh in the late 1990s. Her work showed that clay formation in riverine weathering environments acts the same way as it does in the oceans, driving river waters isotopically heavy (Huh et al., 2001; Huh et al., 1998). This isotopic fractionation, combined with the silicate origin of Li, is the basis for the increasing use of Li isotopes as a tracer of silicate weathering processes.

This chapter will discuss the possibilities for using Li isotopes as a weathering tracer in the modern environment, as well as some of the limitations. It will then examine the potential archives of Li for reconstructing past weathering processes, before moving on to examples of the use of Li in tracing weathering through climatic perturbations in the geologic past, focusing on times prior to the Cenozoic.

Lithium Isotopes as a Tracer of Silicate Weathering Intensity

Dissolved Riverine Fluxes of Lithium and Their Isotopic Composition

Primary silicate rocks have a relatively narrow range in $\delta^7\text{Li}$ (MORB: 3–5‰, continental crust: −10 to +10‰, with a mean of 0.6 ± 0.6‰ (Tomascak et al., 2016)). River waters, on the other hand, have a wide range of 2–44‰ (Dellinger et al., 2015; Huh et al., 1998; Murphy et al., 2019), universally isotopically heavier than the rocks they drain (Fig. 2). It has been demonstrated that carbonate weathering insignificantly affects Li dissolved in rivers, even in carbonate-dominated catchments (Kisakürek et al., 2005). Evaporites can, on occasion, affect local riverine Li isotope signals (Gou et al., 2019). Overall, Li isotopes are, for the most part, a selective tracer of silicate weathering processes, in contrast to isotopic systems such as Sr and Os. Although some Li isotope fractionation has been reported in some plants (Li et al., 2020), other studies report no effect (Clergue et al., 2015; Lemarchand et al., 2010). In all cases the lithium content in vegetation is low, and hence its effect on Li cycles is expected to be relatively small. No effect from phytoplankton on Li isotope ratios has been observed (Pogge von Strandmann et al., 2016), which overall gives Li isotopes an advantage over other many stable isotope tracers including Si, B, Mg and Ca isotopes.

Dissolution of rocks drives the $\delta^7\text{Li}$ of the waters towards the value of the rock (low $\delta^7\text{Li}$, as shown by scenario (i) in Fig. 1). In contrast, secondary mineral formation during weathering (including clays, zeolites and oxyhydroxides) drives water $\delta^7\text{Li}$ to higher values (scenarios (ii), (iii) and (iv) in Fig. 1). This means that the $\delta^7\text{Li}$ value of river waters is determined by the ratio of primary mineral dissolution to secondary mineral formation – a process known

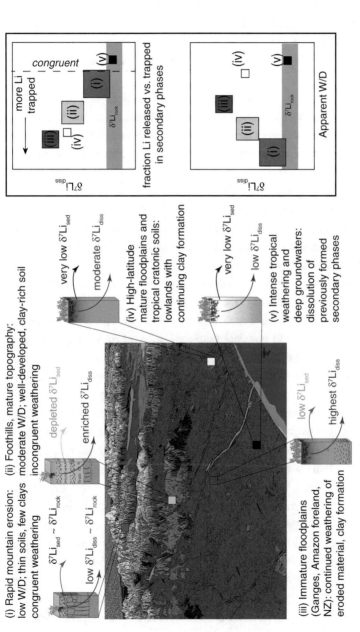

Figure 1 Cartoon showing the processes that fractionate dissolved and sediment δ^7Li during weathering. The different scenarios are detailed in the text. The panels on the right show the positioning of the different scenarios on plots of dissolved δ^7Li as a function of the fraction of Li trapped in secondary phases (top), and as a function of the weathering intensity (W/D) (bottom). The size of the squares represents the dissolved Li yield (e.g. in kg.km^{-2}.yr^{-1}) of the different scenarios.

as the weathering congruency (Misra and Froelich, 2012; Pogge von Strandmann and Henderson, 2015). When weathering is congruent, rocks dissolve with little secondary mineral formation, and water δ^7Li is low (close to the starting rock value). Conversely, when weathering is incongruent, secondary minerals form, and dissolved δ^7Li is high. A final scenario, where previously formed clays redissolve (v), drives δ^7Li low again.

Data from the Amazon River (Dellinger et al., 2015), Ganges River (Pogge von Strandmann et al., 2017b), High Himalayas (Bohlin and Bickle, 2019), New Zealand (Pogge von Strandmann and Henderson, 2015) and Yangtze (Ma et al., 2020) suggest that in areas where the exposure rate of fresh rock is rapid and water–rock contact time is relatively short (e.g. in mountains), riverine δ^7Li values are low and rocklike, as a result of relatively congruent weathering (Fig. 1). In contrast, on flatter floodplains, riverine δ^7Li increases, that is, weathering becomes more incongruent (Bagard et al., 2015; Dellinger et al., 2015; Pogge von Strandmann and Henderson, 2015), probably because there is longer water–sediment interaction time, leading to more clay mineral formation (Liu et al., 2015; Wanner et al., 2014).

These observations lead to the notion that riverine δ^7Li is related to the intensity of silicate weathering (Dellinger et al., 2015), that is, the ratio of the weathering rate (W) to the denudation rate (D). Denudation is the sum of chemical weathering and physical erosion – so weathering intensity effectively gives the ratio of chemical silicate weathering to physical erosion rates. Mountainous areas have high erosion rates, and hence a low W/D value (low weathering intensity) and a low riverine δ^7Li value, but also a high river Li yield (flux per unit area), because there are few clays to remove cations. Due to the lack of secondary mineral formation, river sediments have δ^7Li close to that of the original rock (Fig 1i). As the downstream topography matures, such as in foothills with more developed soils, clays begin to form and water δ^7Li increases, while sediment δ^7Li begins to decrease (Fig. 1ii). Further downstream, immature floodplains have even higher weathering intensity, with relatively greater clay formation relative to rock dissolution, which drives riverine δ^7Li high, the sediment δ^7Li low and the dissolved Li yield lower (Fig. 1iii) (Dellinger et al., 2015; Pogge von Strandmann et al., 2020). High-intensity weathering regimes also exist. These tend to be supply-limited (i.e. little supply of primary rock to dissolve), such as in high-latitude and tropical lowlands (Fig. 1iv), and even more extremely in tropical rainforests (Fig. 1v). As weathering intensity increases, previously formed secondary clays start to be leached. The dissolved δ^7Li signal of these regimes is low again, but the Li yield is also extremely low, suggesting that these supply-limited regimes will have less direct influence on the mass balance of seawater. River sediments have extremely low δ^7Li in these regimes. Combined, these different process regimes yield a "boomerang" shape when plotting dissolved δ^7Li against W/D. Importantly, the dissolved Li yield increases as W/D decreases (Fig. 1 insert; Fig 2a), and hence for the marine mass balance, scenarios (i), (ii) and (iii) are expected to be most relevant.

There is relatively little difference in the riverine δ^7Li ranges between tropical, temperate and polar rivers (Fig. 2) (Murphy et al., 2019), suggesting that the fractionating processes and fractionation factors remain broadly the same, regardless of climate. It is also noticeable that the general "boomerang" shape in the δ^7Li–W/D relationship is observed independently in rivers from different terrains, such as basaltic, polar and tropical rivers (Fig. 2), with higher δ^7Li in intermediate W/D regimes in all cases. Hence, the weathering regime appears to be the dominant control on the dissolved Li isotope ratio, and not lithology or climate.

Ocean Processes Affecting Lithium Isotopes

Lithium in modern oceans has a residence time of approximately 1 million years (Huh et al., 1998). Rivers make up ~60 per cent of the modern ocean inputs (~8×10^9 mol/yr) (Hathorne and James, 2006; Huh et al., 1998). The other major input is from high-temperature hydrothermal fluids at mid-ocean ridges (MORs), which, if modern oceans are at steady state, is ~6×10^9 mol/yr. The Li sinks from seawater are uptake onto low-temperature clays that form during alteration of oceanic crust (AOC) and during the formation of marine authigenic aluminous clays (MAAC) (Chan et al., 1992; Hathorne and James, 2006; Misra and Froelich, 2012). The mean δ^7Li of the global riverine dissolved flux is ~23‰ (Huh et al., 1998), and the average hydrothermal input is ~8‰ (Penniston-Dorland et al., 2017). Combined, this would give seawater a δ^7Li value of ~17‰. However, the sink into clay minerals imparts a combined fractionation factor of 14–16‰, driving the modern seawater value to 31.2 ± 0.2‰ (Fig. 3a) (Jeffcoate et al., 2004).

It can be assumed that the hydrothermal δ^7Li value has remained constant through time, because it is effectively basalt weathering at high temperatures, which is characterized by low fractionation (Fig. 3b). If the hydrothermal flux through time can be extrapolated from MOR spreading rates, then rivers remain the primary unknown. In high-weathering-intensity regimes, the Li flux is low, and hence such a regime globally would mean that the hydrothermal input would dominate the seawater Li budget rather than the river δ^7Li directly (e.g. Seyedali et al., 2021). When weathering fluxes globally are dominated by lower-intensity weathering, as is the case in the present day, river δ^7Li will have a direct influence on seawater δ^7Li, and it appears that the river flux and isotope ratio are linked: when the isotope ratio is low (congruent weathering), the flux is high, while when there is more clay formation (incongruent weathering), the flux decreases and the isotope ratio increases (Fig. 1) (Dellinger et al., 2015). Hence, it may be possible to use this relationship to

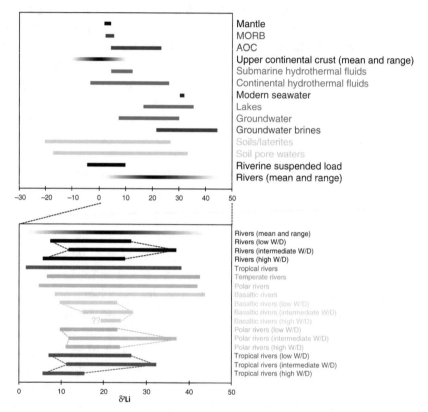

Figure 2 Ranges of δ^7Li of different solid and dissolved Li reservoirs (Penniston-Dorland et al., 2017). MORB = Mid-Ocean Ridge Basalt; AOC = Altered Oceanic Crust. The expanded plot shows the δ^7Li ranges for different river scenarios (Clergue et al., 2015; Dellinger et al., 2017; Dellinger et al., 2015; Gou et al., 2019; Huh et al., 2001; Huh et al., 1998; Kisakürek et al., 2005; Ma et al., 2020; Millot et al., 2010; Murphy et al., 2019; Pogge von Strandmann et al., 2016; Pogge von Strandmann et al., 2017b; Pogge von Strandmann and Henderson, 2015; Tomascak et al., 2016). It is noticeable that for each detailed regime (tropical, polar and basaltic rivers), intermediate W/D rivers range to higher δ^7Li values than low or high W/D rivers. We note that very few basaltic rivers exist with high W/D, and hence this is denoted with '??'.

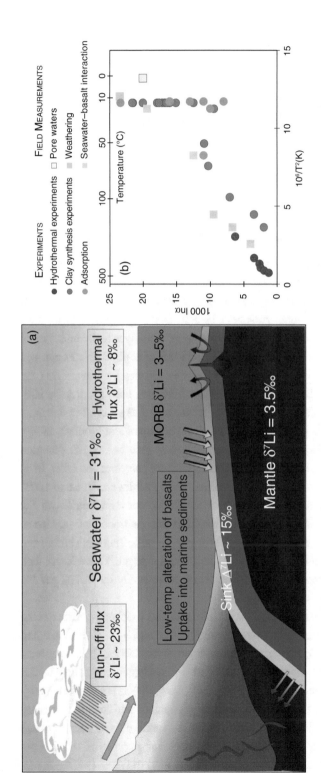

Figure 3 (a) The seawater mass balance of Li, showing the primary sources and sinks, and their isotope ratios. (b) The temperature dependence of Li isotope fractionation, based on experiments and field measurements (Hindshaw et al., 2019; Li and West, 2014; Li and Liu, 2020; Pistiner and Henderson, 2003; Pogge von Strandmann et al., 2019a; Wimpenny et al., 2015; Wimpenny et al., 2010). The weathering data point is based on the global riverine mean (Huh et al., 1998).

extrapolate past changes in weathering flux and processes from the seawater isotope composition.

Because of the temperature dependence of the fractionation factor (Li and West, 2014; Vigier et al., 2015), the fractionation factors associated with AOC and MAAC are different. Therefore 'sink-shifts' that change the proportion of AOC and MAAC (i.e. changes in global fluxes of reverse weathering in marine sediments or weathering of seafloor basalts) can also potentially alter the seawater isotope ratio (Coogan et al., 2017; Li and West, 2014).

Methods and Archive Materials

Sample Preparation and Mass Spectrometry

In early studies, Li isotope ratios were measured by thermal ionization mass spectrometry (TIMS) (Chan et al., 1992). However, high-precision TIMS analyses of a system with only two stable isotopes are challenging. Hence, for the past 25 years, Li isotope ratios have been analysed by multi-collector ICP-MS. The latest generation of mass spectrometers can analyse less than 0.5 ng of Li to a high precision (Bohlin et al., 2018; Jeffcoate et al., 2004; Pogge von Strandmann et al., 2019a), which allows analyses of almost all materials (e.g. carbonates, waters, silicates, plants). Some studies have used quadrupole ICP-MS because this requires less sample on the whole (0.2–2 ng Li), but precision is considerably worse (<±1.5‰)(Liu and Li, 2019; Misra and Froelich, 2009), compared to <±0.5‰ by multi-collector ICP-MS (see below). Purification chemistry involves cation exchange columns, where care must be taken to 1) remove all matrix from the purified solution, and 2) achieve as close to 100% column yields as possible. This is because Li isotopes fractionate on the columns, so low yields will cause unquantifiable isotope fractionation. Hence, yields must always be monitored for each sample, by collecting elute before and after the Li collection bracket and analysing for Li content. Yields lower than 100% will cause reproducibility to worsen, observationally by ~1.7‰ per 1% loss of yield (Wilson, pers. comm).

Isotopic analysis is conducted by sample-standard bracketing, either using the L-SVEC standard, or the newer, virtually identical, IRMM-016 standard (Jeffcoate et al., 2004; Pogge von Strandmann et al., 2019a). The long-term external error should always be quantified, by analysing the same natural standard (e.g. seawater, or international rock standards) every session. Long-term analysis (i.e. up to 50 or 100 separate sessions of purification chemistry and analysis) can yield 2sd errors of <0.5‰ on modern mass spectrometers (reported down to ±0.3‰ on some machines over periods of >five years (Jeffcoate et al., 2004; Pogge von Strandmann et al., 2011)).

In terms of examining Li isotope behaviour in the geological record, two types of archive have been used. The main one is marine carbonates: bulk, foraminifera, brachiopods, belemnites, corals and bivalves. Inorganic carbonates (speleothems) also have been used. The second, less well-established archive is clays, either detrital or authigenic.

Carbonates

Different biogenic carbonates (e.g. foraminifera, brachiopods) have different partition coefficients for Li that are also different from those for inorganic carbonates (Dellinger et al., 2018). It has also been shown that the uptake of Li into inorganic carbonates is strongly temperature-dependent and also may be affected by the solute concentration, ionic strength and pH (Marriott et al., 2004a; Marriott et al., 2004b). For this reason, although studies typically report Li/Ca ratios, they are rarely used for interpretational purposes.

In contrast, some carbonates exhibit narrow ranges in Li isotope fractionation (Dellinger et al., 2018). Modern bulk aragonites have a fractionation factor of $\Delta^7Li_{aragonite-seawater}$ ~−10‰ (Gabitov et al., 2011; Marriott et al., 2004b; Pogge von Strandmann et al., 2019b), in agreement with inorganic aragonites (Marriott et al., 2004b; Pogge von Strandmann et al., 2019b). Similarly, biogenic aragonites such as corals (mean $\Delta^7Li_{aragonite-seawater}$ = −12±2‰, 1σ, (Bastian et al., 2018; Marriott et al., 2004a; Marriott et al., 2004b; Rollion-Bard et al., 2009) and aragonitic mollusks (mean $\Delta^7Li_{aragonite-seawater}$ = −13±3‰, 1σ (Bastian et al., 2018; Dellinger et al., 2020)) have Δ^7Li similar to inorganic aragonite (~−10‰). Calcites tend to have lower fractionation factors than aragonite, with $\Delta^7Li_{calcite-seawater}$ ~−6‰ in modern core tops. High-Mg calcites have fractionation factors between the two (Dellinger et al., 2018).

Bulk carbonates, foraminifera and brachiopods (and potentially belemnites) appear to exhibit a fairly narrow range in fractionation factor (Dellinger et al., 2018; Ullmann et al., 2013), although there have been contrasting reports on foraminifera (Fig. 4). Planktic foraminifera have δ^7Li closer to seawater (i.e. less negative $\Delta^7Li_{calcite-seawater}$) compared to benthic foraminifera (Dellinger et al., 2018; Roberts et al., 2018), and there appear to be species-specific vital effects (Hathorne and James, 2006), although this is disputed (Misra and Froelich, 2012). There have also been reports of pH effects (Roberts et al., 2018), although again this is disputed (Vigier et al., 2015). The conflicting interpretations suggest that there may be complications associated with using foraminifera as a Li isotope archive – although Cenozoic δ^7Li records from foraminifera have been independently replicated in brachiopods (Misra and Froelich, 2012; Washington et al., 2020), and in-depth studies similar to those

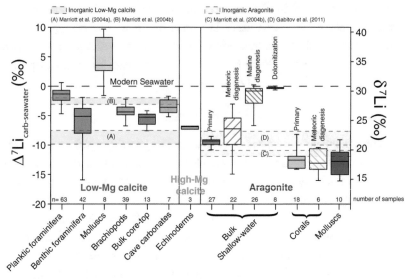

Figure 4 Plot of δ⁷Li and (Δ⁷Li $_{carb-seawater}$ = δ⁷Li$_{carbonate}$ − δ⁷Li$_{seawater}$) ranges in modern carbonates, including the effect of diagenetic recrystallization of aragonite to low-Mg calcite and dolomite (see text for references). The boxes reflect the interquartile range with the median shown as the line and the extreme data points as the whiskers. The number of samples is indicated at the bottom of the figure. For biogenic carbonates, the number of samples corresponds to the number of different specimens analysed.

on foraminifera will need to be conducted on other types of biogenic carbonates. Bulk carbonates appear to be fairly robust archives of seawater δ⁷Li (Dellinger et al., 2020; Pogge von Strandmann et al., 2019b), and within such carbonates the proportion of foraminifera to coccoliths does not affect the overall δ⁷Li value, suggesting that both have similar overall fractionation factors (Pogge von Strandmann et al., 2019b). Finally, it appears that there are no temperature effects on the Li isotope fractionation factor in either inorganic or biogenic calcites (Dellinger et al., 2018). In contrast, molluscs appear to exhibit highly variable fractionation (on occasion, even in different layers from the same shells) and may not be so useful as a Li isotope archive (Dellinger et al., 2018).

Carbonate diagenesis can preserve or change the primary δ⁷Li of carbonate depending upon the composition of the original carbonate (aragonite or calcite) and of the diagenetic fluid (marine or meteoric), and also depending on diagenesis style (fluid or sediment buffered) (Dellinger et al., 2020). Early marine diagenetic recrystallization of aragonite to calcite under fluid-buffered conditions (typical of platform top margins), as well as dolomitization, result in a narrow range of Li isotope compositions that is similar to corresponding seawater at the time of diagenesis ($\Delta^7Li_{calcite-seawater}$ ∼−1‰) but distinct from the original aragonite ($\Delta^7Li_{aragonite-seawater}$∼−10‰) (Fig. 4). In contrast, marine diagenesis of aragonite under a continuum of fluid- to sediment-buffered conditions (such as on platform slopes) imparts a larger isotopic variability ($\Delta^7Li_{calcite-seawater}$ between −1 and −10‰). Meteoric diagenesis of platform margins under fluid-buffered conditions overprints the primary aragonite composition with a highly variable isotope signature ($\Delta^7Li_{calcite-seawater}$ of −3 to −12‰), although the average composition can be coincidentally similar to the original (Dellinger et al., 2020). Diagenesis of biogenic low-Mg calcite (as evidenced by increasing Mn/Ca) has been observed to produce a decrease of δ^7Li of 2 to 7‰ for planktic forami- nifera (Bastian et al., 2018) and increase of up to 13‰ for belemnite rostra (Ullmann et al., 2013). Altogether, these studies show that some diagenetic carbon- ates can faithfully record past seawater composition, provided that their diagenetic history can be independently constrained (e.g. using Sr/Ca, Mn/Ca or Ca isotopes).

Clays

In principle, clays can also be an archive of fluid Li isotope ratios. Marine authigenic clays (either AOC or MAAC) should record seawater δ^7Li with a constant (albeit temperature-dependent) fractionation factor. However, as yet this archive has not been tested. In contrast, several studies have measured Li isotopes in detrital clays, as an archive of local weathering conditions (Bastian et al., 2017; Li et al., 2016; Pogge von Strandmann et al., 2017a; Wei et al., 2020). Detrital secondary minerals should form predominantly in soils, and hence these records should reflect the local weathering conditions. Detrital material can also consist of partially weathered primary minerals and clays derived from the erosion of sedimentary rocks. Modern river data suggest that the isotopic difference between source rock and clay fraction decreases with increasing weathering intensity (Dellinger et al., 2017), partly because a larger proportion of the total transported Li is in sediments at higher W/D, and possibly because primary material is less altered at low W/D (Fig. 1).

There are two principal problems that currently prevent full exploitation of this archive: (1) clay fractionation factors are not well known, so absolute water δ^7Li values cannot be fully reconstructed, and (2) a change in the source of weathering can change the isotope ratio of the rock being weathered and eroded, and potentially the clay mineral type being formed. However, some of these problems can be alleviated by also analysing radiogenic isotopes or other geochemical indices as source tracers, along with characterizing the clay min- eralogy of the detrital material.

Li Isotopes in Pre-Cenozoic Archives

Records of past δ^7Li have been generated at relatively low resolution over the Cenozoic (Hathorne and James, 2006; Misra and Froelich, 2012), at several individual points from the Archean and Proterozoic based on detrital material (Li et al., 2016), and at higher resolution across several short climate perturbations, such as Oceanic Anoxic Events (OAEs) (e.g. Pogge von Strandmann et al., 2013).

This Element focuses on pre-Cenozoic records, so we merely note that while the Cenozoic increases in seawater δ^7Li have been observed in foraminifera (Hathorne and James, 2006; Misra and Froelich, 2012) and reproduced in brachiopods (Washington et al., 2020), the interpretation of this change remains highly controversial (Caves Rugenstein et al., 2019; Li and West, 2014; Vigier and Godderis, 2015; Wanner et al., 2014). This is largely because the Cenozoic is itself a complex time period with substantial long-term changes of both tectonic and climatic conditions, adding complexity to the interaction of the weathering cycle with climate change.

Numerous studies have examined short, rapid climatic perturbations for Li isotopes, which provide more direct evidence for the response of global weathering to climatic change. Lithium isotope records across such intervals include carbonate records from the Hirnantian glaciation (Pogge von Strandmann et al., 2017a) and OAE 1a and 2 (Lechler et al., 2015; Pogge von Strandmann et al., 2013). Detrital records have been published from the Sturtian deglaciation (Wei et al., 2020) and the Hirnantian glaciation (Pogge von Strandmann et al., 2017a). Data from single time points have been reported from Jurassic carbonates (Ullmann et al., 2013) and throughout the Proterozoic and early Phanerozoic, from glacial diamictites (Li et al., 2016). A bulk sedimentary rock record has also been presented from the Permo-Triassic boundary (Sun et al., 2018); however, it is not entirely clear how well this record reflects seawater Li isotope ratios (as discussed below), because the carbonate δ^7Li was extrapolated from bulk sediment (including silicate) δ^7Li, using a mixing model assuming constant fractionation factors.

Carbonate Records of Seawater δ^7Li across Major Climatic Perturbations

A compilation of the carbonate δ^7Li records associated with climate perturbations in the past shows apparently reproducible behaviour of seawater δ^7Li in response to climatic change (Fig. 5), also reproduced in speleothem records of local solution δ^7Li (Pogge von Strandmann et al., 2017c). During warming episodes, most notably associated with the Cretaceous OAEs, carbonate δ^7Li exhibits a negative excursion. Simple dynamic box models of Li isotope behaviour in seawater show that a transient negative excursion

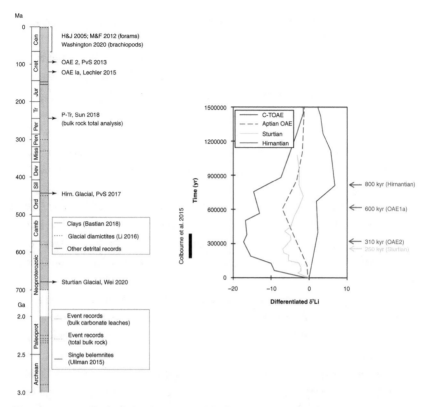

Figure 5 A geological timeline, showing the Li isotope studies published to date and the type of archive used (Lechler et al., 2015; Pogge von Strandmann et al., 2017a; Pogge von Strandmann et al., 2013; Wei et al., 2020). We focus on four Pre-Cenozoic records (three warming periods and one cooling period) and show their measured differentiated δ^7Li values against time (a number is subtracted from the δ^7Li values of each section so that they all start at 0; t = 0 is selected as the point at which δ^7Li deviates from the pre-existing steady state, and the individual age models are taken from each study). The inflection points represent the likely time (given on the right) it took for weathering to influence climate, thereby demonstrating the weathering thermostat. Timescales of the silicate weathering feedback from modelling studies are shown on the left (Colbourn et al., 2015).

in seawater can be caused by either an increase in the riverine flux or a decrease in the riverine δ^7Li values (Pogge von Strandmann et al., 2013). However, in most cases, one of these changes by itself is not sufficient to explain the observed records, because riverine δ^7Li is unlikely to decrease below the value of the continental crust (~0‰), and extreme increases in the riverine flux must be reconciled with other records, such as from radiogenic Sr isotopes. As in modern rivers (Dellinger et al., 2015), it is likely that river flux and δ^7Li are coupled (Fig. 1). A combined increase in flux and decrease in δ^7Li can drive large-magnitude transient changes in seawater δ^7Li values, consistent with those observed. In principle, an increase in the hydrothermal flux can also drive a negative seawater δ^7Li excursion. However, in general, the necessary increase in flux is of an unfeasible size and would produce larger changes in the radiogenic Sr or Os isotope records than observed in these events (Pogge von Strandmann et al., 2013).

Mirroring the observations of negative seawater δ^7Li excursions during warming events, carbonate records show a positive δ^7Li excursion during cooling episodes (Pogge von Strandmann et al., 2017a). These have been interpreted as a decrease in the riverine flux, coupled with an increase in riverine δ^7Li values.

When interpreting seawater δ^7Li records, it must be noted that while carbonate isotopic fractionation factors are not temperature-dependent, those of clay formation are (Li and West, 2014). Hence, an increase in temperature would cause lower riverine δ^7Li values, even without an associated change in weathering rates or regimes. This temperature dependence is approximately 0.17‰/°C (Gou et al., 2019; Pogge von Strandmann et al., 2017a), so while the direction of fractionation is correct for parts of the Li isotope records to be a temperature response of the clay fractionation factor, in almost all cases the change is not sufficient to drive the observed excursions in isolation. The exception is potentially the P-Tr record (Sun et al., 2018), where up to 100 per cent of the ~14‰ excursion could be driven by temperature-dependent fractionation factors. (We also note that the excursion recovery of <200 kyr in this study would require a Li ocean residence time an order of magnitude below that observed in other studies). In contrast, ~5% of the record at OAE2 could be temperature-dependent, 25% of the record at OAE1a, and ~15% of the record at the Hirnantian.

Interpreting Detrital δ^7Li Records from Deep Time

The interpretation of changes in weathering intensity from detrital records in the pre-Cenozoic has proven somewhat more complicated. For example, after the Sturtian Snowball Earth event, a negative detrital δ^7Li excursion is interpreted as reflecting an increase in silicate weathering intensity (Wei et al., 2020), while at the Hirnantian glaciation, a positive δ^7Li excursion is also interpreted as an increase in weathering intensity (Pogge von Strandmann et al., 2017a). The reason for this interpretational ambiguity is partly because detrital material can contain both altered primary material and secondary material (Dellinger et al., 2017) as well as clays formed during MAAC. Also, modern detrital samples exhibit a relatively high amount of scatter, which means that changes observed in the geological record can disappear in the noise. More work is needed to understand what determines the composition of detrital material, in particular what determines the clay–water δ^7Li fractionation factor. Regardless of this current interpretational ambiguity, the fact remains that, like carbonate records, warming has led to a negative detrital δ^7Li excursion, while cooling has led to a positive excursion – suggesting that these records offer promising opportunities.

Seawater Li Isotope Records and the Global Weathering Thermostat

Overall, the event-scale Li isotope records appear to confirm the operation of the 'weathering thermostat', that is, a dominant temperature control on the weathering rate, affecting the CO_2 removal rate, thus stabilizing climate (Fig. 5). The Li isotope records suggest that weathering rates increase during warming and decrease during cooling. At the same time, increases in temperature appear to shift the weathering intensity towards lower values, and vice versa for decreasing temperatures (Lechler et al., 2015; Pogge von Strandmann et al., 2017a; Pogge von Strandmann et al., 2013). This broadly suggests that as temperatures increase, at least in transient perturbations, erosion rates increase more than weathering rates. This is potentially counter-intuitive, given that temperature is generally thought to increase weathering rates. More work is needed, and we note that the warming episodes studied were caused by large basaltic eruptions (Pogge von Strandmann et al., 2013), likely increasing the supply of fresh material. It may therefore be that in the geological record of such events, the 'D' component of weathering intensity can be affected not only by erosion but also by the supply of fresh rock by volcanism.

The high-resolution records of seawater δ^7Li values across climatic perturbations allow testing of the rate of response of the weathering thermostat. The inflection point of each record should represent the time by which weathering increased to its maximum (in the case of warming) or decreased to its minimum (in the case of cooling), and hence the time period by which weathering effectively gained control of the climate (Fig. 5 – using the age models reported in each study). Records studied to date have inflection points between 250 and 800 kyr from inception (250–600 kyr for warming events). These timescales are similar to those determined in models of weathering in warming climates (200–

400 kyr after inception (Colbourn et al., 2015). Hence, models and the real world, as inferred from Li isotopes, apparently give similar values, albeit with more scatter in the isotopic records – likely due to continental placement, the supply of rock available for weathering (i.e. terrain), plus, in the case of warming, the duration and pattern of carbon outgassing. The initial Li budget of the oceans before any perturbation will also affect buffering of the Li response time.

Developing a Mechanistic Understanding of Li Isotope Fractionation

This section has described the basic qualitative framework, now fairly well understood, for the response of the Li isotope system to changes in weathering. This response appears robust across wide-ranging modern environments, and past Li isotope records through climate perturbations yield consistent results that are in line with independent expectations (e.g. from carbon cycle models). Yet a major piece that is still missing in the robust development and application of this proxy is a mechanistic understanding of the fractionation taking place during weathering and in the oceans. Reactive transport models that quantitatively link isotopic fractionation with fluid and mineral residence times show promise in this regard (e.g. Bohlin and Bickle, 2019; Wanner et al., 2014). Yet applying such models to interpret geological records remains challenging.

One of the major challenges to applying mechanistic models lies in the limited understanding of Li isotope fractionation factors. Laboratory experiments including direct mineral synthesis (Hindshaw et al., 2019; Vigier et al., 2008), water–rock alteration of primary minerals (Pogge von Strandmann et al., 2019a; Wimpenny et al., 2010), and measurements of exchangeable lithium (Li and Liu, 2020; Pistiner and Henderson, 2003; Wimpenny et al., 2015; Wimpenny et al., 2010) have confirmed field observations that the light Li isotope preferentially goes into alteration phases. Importantly, the experimental work has revealed that there are two sites that can take up Li in clays: (1) interlayer sites within forming minerals (i.e. structurally bound Li during neoformation), and (2) exchangeable sites on mineral surfaces (i.e. adsorbed Li; Vigier et al., 2008). While both site types prefer ^{6}Li, the fractionation factors are thought to be different (Hindshaw et al., 2019). Fractionation on exchangeable sites appears smaller (0–12‰) than on structural sites (14–24‰). Natural samples will exhibit fractionation that is a balance of structural and exchangeable Li – lending complexity to the extrapolation from lab-derived fractionation factors to geological settings.

The dependence of fractionation on environmental conditions may also be important. There is good evidence that temperature exerts a control, as discussed earlier (Fig. 3b). However, there is no experimental evidence that the fractionation factors depend on parameters such as pH or solution composition (Hindshaw et al., 2019). It is still disputed whether the identity of the reacting mineral or the secondary minerals undergoing formation can affect the fractionation factor (Hindshaw et al., 2019). The types of clay that can be experimentally synthesized are extremely limited, so there is no clear evidence either way for structural sites. For exchangeable sites, there is contrasting information from different studies (Hindshaw et al., 2019; Pistiner and Henderson, 2003; Pogge von Strandmann et al., 2019a; Wimpenny et al., 2015; Wimpenny et al., 2010), and more work is needed.

There is considerable uncertainty in our knowledge not only of fractionation factors but also of Li partition coefficients into different secondary minerals (Decarreau et al., 2012). The latter is almost more important, because it is not necessarily known which secondary minerals have a large influence on the Li weathering budget (e.g. if oxyhydroxides have a partition coefficient similar to that of clays, they also need experimental focus). Overall, accurate fractionation factors and partition coefficients are perhaps the key hurdle for Li isotopes to move from a qualitative to a quantitative weathering tracer.

Future Prospects

Lithium isotopes, in both carbonates and clays, have significant potential for reconstructing and understanding weathering and hence carbon cycle behaviour throughout Earth's history. The primary advancement needed to use Li isotopes quantifiably is a better knowledge of fractionation factors and partitioning behaviour during secondary mineral formation. This must perforce be performed experimentally but need not necessarily take the form of measuring fractionation factors for individual secondary minerals, because single types of secondary mineral rarely form in the natural environment. One alternative is to artificially weather different rock types and measure fractionation factors of the resulting conglomerations of multiple secondary minerals. Improved understanding of fractionation, combined with a full mechanistic link between riverine flux and isotope ratio in different environments, would allow a quantitative determination of weathering behaviour. Models that link Li isotopes to the carbon cycle already have been presented (Caves Rugenstein et al., 2019; Li and West, 2014; Pogge von Strandmann et al., 2017a; Vigier and Godderis, 2015), but more natural and experimental details are needed to clear the ambiguity between these.

As a final point, while much of the attention in this section has focused on the way that Li isotopes reflect continental weathering processes, it is important to recognize that seawater δ^7Li values are also sensitive to past marine authigenic clay formation (i.e. reverse weathering (Li and West, 2014), as well as to seafloor weathering (Coogan et al., 2017)). These processes can also have significant effects on the carbon cycle and are currently poorly constrained. Given the significant isotope fractionation that reverse weathering causes to Li isotopes, they may provide an exploitable proxy over various timescales. Yet, application to understanding these processes will depend on distinguishing their effects from those of changes in continental weathering fluxes and intensity.

References

Key Papers

1. Chan, L. H., Edmond, J. M., Thompson, G., and Gillis, K. (1992). Lithium isotopic composition of submarine basalts: Implications for the lithium cycle in the oceans. *Earth Planet. Sci. Lett.* 108, 151–160. – First demonstration of low-temperature Li isotope fractionation, during alteration of the oceanic crust.

2. Huh, Y., Chan, L. H., Zhang, L., and Edmond, J. M. (1998). Lithium and its isotopes in major world rivers: Implications for weathering and the oceanic budget. *Geochim. Cosmochim. Acta* 62, 2039–2051. – First compilation of global riverine δ^7Li.

3. Pistiner, J. S., and Henderson, G. M. (2003). Lithium-isotope fractionation during continental weathering processes. *Earth Planet. Sci. Lett.* 214, 327–339. – First experimental examination of Li isotope behaviour during weathering.

4. Kisakürek, B., James, R. H., and Harris, N. B. W. (2005). Li and δ^7Li in Himalayan rivers: Proxies for silicate weathering? *Earth Planet. Sci. Lett.* 237, 387–401. – First demonstration that Li isotopes track silicate weathering only.

5. Misra, S., and Froelich, P. N. (2012). Lithium isotope history of Cenozoic seawater: Changes in silicate weathering and reverse weathering. *Science* 335, 818–823. – Record of the seawater of the entire Cenozoic.

6. Pogge von Strandmann, P. A. E., Jenkyns, H. C., and Woodfine, R. G. (2013). Lithium isotope evidence for enhanced weathering during Oceanic Anoxic Event 2. *Nature Geoscience* 6, 668–672. – First short-term record through a geological event.

7. Dellinger, M., Gaillardet, J., Bouchez, J., et al. (2015). Riverine Li isotope fractionation in the Amazon River basin controlled by the weathering regimes. *Geochim. Cosmochim. Acta* 164, 71–93. – Paper that demonstrates the riverine Li isotope response to weathering intensity.

8. Pogge von Strandmann, P. A. E., and Henderson, G. M. (2015). The Li isotope response to mountain uplift. *Geology* 43, 67–70. – Demonstration of impact of supply vs. residence time on Li isotopes.

9. Dellinger, M., Hardisty, D. S., Planavsky, N. J., et al. (2020). The effects of diagenesis on lithium isotope ratios of shallow marine carbonates. *Am. J. Sci.* 320, 150–184. – Carbonate fractionation factors, and their use for palaeo-record reconstruction.

10. Hindshaw, R. S., Tosca, R., Gout, T. L., et al. (2019). Experimental constraints on Li isotope fractionation during clay formation. *Geochim. Cosmochim. Acta* 250, 219–237 – Fractionation during clay synthesis.

Other References

Bagard, M.-L., West, A. J., Newman, K., and Basu, A. R. (2015). Lithium isotope fractionation in the Ganges–Brahmaputra floodplain and implications for groundwater impact on seawater isotopic composition. *Earth Planet. Sci. Lett.* 432, 404–414.

Bastian, L., Revel, M., Bayon, G., Dufour, A., and Vigier, N. (2017). Abrupt response of chemical weathering to Late Quaternary hydroclimate changes in northeast Africa. *Scientific Reports* 7, 44231.

Bastian, L., Vigier, N., Reynaud, S., et al. (2018). Lithium isotope composition of marine biogenic carbonates and related reference materials. *Geostandards and Geoanalytical Research* 23, 403–415.

Bohlin, M. S., and Bickle, M. J. (2019). The reactive transport of Li as a monitor of weathering processes in kinetically limited weathering regimes. *Earth Planet. Sci. Lett.* 511, 233–243.

Bohlin, M. S., Misra, S., Lloyd, N., Elderfield, H., and Bickle, M. J. (2018). High-precision determination of lithium and magnesium isotopes utilising single column separation and multi-collector inductively coupled plasma mass spectrometry. *Rapid Communications in Mass Spectrometry* 32, 93–104.

Caves Rugenstein, J. K., Ibarra, D. E., and von Blanckenburg, F. (2019). Neogene cooling driven by land surface reactivity rather than increased weathering fluxes. *Nature* 571, 99–102.

Clergue, C., Dellinger, M., Buss, H. L., et al. (2015). Influence of atmospheric deposits and secondary minerals on Li isotopes budget in a highly weathered catchment, Guadeloupe (Lesser Antilles). *Chem. Geol.* 414, 28–41.

Colbourn, G., Ridgwell, A., and Lenton, T. M. (2015). The time scale of the silicate weathering negative feedback on atmospheric CO_2. *Global Biogeochemical Cycles* 29, 583–596.

Coogan, L. A., Gillis, K. M., Pope, M., and Spence, J. (2017). The role of low-temperature (off-axis) alteration of the oceanic crust in the global Li-cycle: Insights from the Troodos ophiolite. *Geochim. Cosmochim. Acta* 203, 201–215.

Decarreau, A., Vigier, N., Pálková, H., et al. (2012). Partitioning of lithium between smectite and solution: An experimental approach. *Geochim. Cosmochim. Acta* 85, 314–325.

Dellinger, M., Bouchez, J., Gaillardet, J., Faure, L., and Moureau, J. (2017). Tracing weathering regimes using the lithium isotope composition of detrital sediments. *Geology* 45 (5). pp. 411–414.

Dellinger, M., West, A. J., Paris, G., et al. (2018). The Li isotope composition of marine biogenic carbonates: Patterns and mechanisms. *Geochim. Cosmochim. Acta* 236, pp. 315–335.

Gabitov, R. I., Schmitt, A. K., Rosner, M., et al. (2011). In situ δ^7Li, Li/Ca, and Mg/Ca analyses of synthetic aragonites. *Geochem. Geophys. Geosyst.* 12, Q03001, http://doi.org/10.1029/2010GC003322.

Gou, L.-F., Jin, Z., Pogge von Strandmann, P. A. E., et al. (2019). Li isotopes in the middle Yellow River: Seasonal variability, sources and fractionation. *Geochim. Cosmochim. Acta* 248, 88–108.

Hathorne, E. C., and James, R. H. (2006). Temporal record of lithium in seawater: A tracer for silicate weathering? *Earth Planet. Sci. Lett.* 246, 393–406.

Huh, Y., Chan, L. H., and Edmond, J. M. (2001). Lithium isotopes as a probe of weathering processes: Orinoco River. *Earth Planet. Sci. Lett.* 194, 189–199.

Jeffcoate, A. B., Elliott, T., Thomas, A., and Bouman, C. (2004). Precise, small sample size determinations of lithium isotopic compositions of geological reference materials and modern seawater by MC-ICP-MS. *Geostandards and Geoanalytical Research* 28, 161–172.

Kennedy, M. J., and Wagner, T. (2011). Clay mineral continental amplifier for marine carbon sequestration in a greenhouse ocean. *Proceedings of the National Academy of Sciences* 108, 9776–9781.

Lechler, M., Pogge von Strandmann, P. A. E., Jenkyns, H. C., Prosser, G., and Parente, M. (2015). Lithium-isotope evidence for enhanced silicate weathering during OAE 1a (Early Aptian Selli event). *Earth Planet. Sci. Lett.* 432, 210–222.

Lemarchand, E., Chabaux, F., Vigier, N., Millot, R., and Pierret, M. C. (2010). Lithium isotope systematics in a forested granitic catchment (Strengbach, Vosges Mountains, France). *Geochim. Cosmochim. Acta* 74, 4612–4628.

Li, G., West, A. J. 2014. Evolution of Cenozoic seawater lithium isotopes: Coupling of global denudation regime and shifting seawater sinks. *Earth Planet. Sci. Lett.* 401, 284–293.

Li, S., Gaschnig, R. M., and Rudnick, R. L. (2016). Insights into chemical weathering of the upper continental crust from the geochemistry of ancient glacial diamictites. *Geochim. Cosmochim. Acta* 176, 96–117.

Li, W., and Liu, X.-M. (2020). Experimental investigation of lithium isotope fractionation during kaolinite adsorption: Implications for chemical weathering. *Geochim. Cosmochim. Acta* 284, 156–172.

Li, W., Liu, X.-M., and Chadwick, O. A. (2020). Lithium isotope behavior in Hawaiian regoliths: Soil-atmosphere-biosphere exchanges. *Geochim. Cosmochim. Acta* 285, 175–192.

Liu, X.-M., and Li, W. (2019). Lithium isotopic analysis by quadrupole-ICP-MS: Optimization for geological samples. *J. Anal. At. Spectrom.* 34, 1706–1717.

Liu, X.-M., Wanner, C., Rudnick, R. L., and McDonough, W. F. (2015). Processes controlling δ^7Li in rivers illuminated by study of streams and groundwaters draining basalts. *Earth Planet. Sci. Lett.* 409, 212–224.

Ma, T., Weynell, M., Li, S. L., et al. (2020). Lithium isotope compositions of the Yangtze River headwaters: Weathering in high-relief catchments. *Geochim. Cosmochim. Acta* 280, 46–65.

Marriott, C. S., Henderson, G. M., Belshaw, N. S., and Tudhope, A. W. (2004a). Temperature dependence of δ^7Li, $\delta^{44}Ca$ and Li/Ca during growth of calcium carbonate. *Earth Planet. Sci. Lett.* 222, 615–624.

Marriott, C. S., Henderson, G. M., Crompton, R., Staubwasser, M., and Shaw, S. (2004b). Effect of mineralogy, salinity, and temperature on Li/Ca and Li isotope composition of calcium carbonate. *Chem. Geol.* 212, 5–15.

Millot, R., Vigier, N., and Gaillardet, J. (2010). Behaviour of lithium and its isotopes during weathering in the Mackenzie Basin, Canada. *Geochim. Cosmochim. Acta* 74, 3897–3912.

Misra, S., and Froelich, P. N. (2009). Measurement of lithium isotope ratios by quadrupole-ICP-MS: Application to seawater and natural carbonates. *J. Anal. At. Spectrom.* 24, 1524–1533.

Murphy, M. J., Porcelli, D., Pogge von Strandmann, P. A. E., et al. (2019). Tracing silicate weathering processes in the permafrost-dominated Lena River water-shed using lithium isotopes. *Geochim. Cosmochim. Acta* 245, 154–171.

Penniston-Dorland, S., Liu, X.-M., and Rudnick, R. L. (2017). Lithium isotope geochemistry, *Rev. Min. Geochem.*, pp. 165–217.

Pogge von Strandmann, P. A. E., Burton,K. W., Opfergelt, S., et al. (2016). The effect of hydrothermal spring weathering processes and primary productivity on lithium isotopes: Lake Myvatn, *Iceland Chem. Geol.* 445, 4–13.

Pogge von Strandmann, P. A. E., Desrochers, A., Murphy, M. J., et al. (2017a). Global climate stabilisation by chemical weathering during the Hirnantian glaciation. *GPL* 3, 230–237.

Pogge von Strandmann, P. A. E., Elliott,T., Marschall, H. R., et al. (2011). Variations of Li and Mg isotope ratios in bulk chondrites and mantle xenoliths. *Geochim. Cosmochim. Acta* 75, 5247–5268.

Pogge von Strandmann, P. A. E., Fraser, W. T., Hammond, S. J., et al. (2019a). Experimental determination of Li isotope behaviour during basalt weathering. *Chem. Geol.* 517, 34–43.

Pogge von Strandmann, P. A. E., Frings, P. J., and Murphy, M. J. (2017b). Lithium isotope behaviour during weathering in the Ganges Alluvial Plain. *Geochim. Cosmochim. Acta* 198, 17–31.

Pogge von Strandmann, P. A. E., Kasemann, S. A., and Wimpenny, J. B. (2020). Lithium and lithium isotopes in Earth's surface cycles. *Elements* 16, 253–258.

Pogge von Strandmann, P. A. E., Schmidt, D. N., Planavsky, N. J., et al. (2019b). Assessing bulk carbonates as archives for seawater Li isotope ratios. *Chem. Geol.* 530, 119338.

Pogge von Strandmann, P. A. E., Vaks, A., Bar-Matthews, M., et al. (2017c). Lithium isotopes in speleothems: Temperature-controlled variation in silicate weathering during glacial cycles. *Earth Planet. Sci. Lett.* 469, 64–74.

Raymo, M. E., Ruddiman, W. F., and Froelich, P. N. (1988). Influence of late Cenozoic mountain building on ocean geochemical cycles. *Geology* 16, 649–653.

Roberts, J., Kaczmarek, K., Langer, G., et al. (2018). Lithium isotopic composition of benthic foraminifera: A new proxy for paleo-pH reconstruction. *Geochim. Cosmochim. Acta* 236, 336–350.

Rollion-Bard, C., Vigier, N., Meibom, A., et al. (2009). Effect of environmental conditions and skeletal ultrastructure on the Li isotopic composition of scleractinian corals. *Earth Planet. Sci. Lett.* 286, 63–70.

Seyedali, M., Coogan, L. A., and Gillis, K. M. (2021). Li-isotope exchange during low-temperature alteration of the upper oceanic crust at DSDP Sites 417 and 418. *Geochim. Cosmochim. Acta* 294, 160–173.

Sun, H., Xiao, Y., Gao, Y., et al. (2018). Rapid enhancement of chemical weathering recorded by extremely light seawater lithium isotopes at the Permian–Triassic boundary. *Proceedings of the National Academy of Sciences* 115, 3782–3787.

Teng, F. Z., Li, W. Y., Rudnick, R. L., and Gardner, L. R. (2010). Contrasting lithium and magnesium isotope fractionation during continental weathering. *Earth Planet. Sci. Lett.* 300, 63–71.

Tomascak, P. B., Magna, T., and Dohmen, R. (2016). *Advances in Lithium Isotope Geochemistry*. Springer.

Ullmann, C. V., Campbell, H. J., Frei, R., et al. (2013). Partial diagenetic overprint of Late Jurassic belemnites from New Zealand: Implications for the preservation potential of d7Li values in calcite fossils. *Geochim. Cosmochim. Acta* 120, 80–96.

Vigier, N., Decarreau, A., Millot, R., et al. (2008). Quantifying Li isotope fractionation during smectite formation and implications for the Li cycle. *Geochim. Cosmochim. Acta* 72, 780–792.

Vigier, N., and Godderis, Y. (2015). A new approach for modeling Cenozoic oceanic lithium isotope paleo-variations: The key role of climate. *Climate of the Past* 11, 635–645.

Vigier, N., Rollion-Bard, C., Levenson, Y., and Erez, J. (2015). Lithium isotopes in foraminifera shells as a novel proxy for the ocean dissolved inorganic carbon (DIC). *C. R. Geosci.* 347, 43–51.

Walker, J. C. G., Hays, P. B., and Kasting, J. F. (1981). A negative feedback mechanism for the long-term stabilization of Earth's surface-temperature. *Journal of Geophysical Research – Oceans and Atmospheres* 86, 9776–9782.

Wanner, C., Sonnenthal, E. L., and Liu, X.-M. (2014). Seawater δ7Li: A direct proxy for global CO_2 consumption by continental silicate weathering? *Chem. Geol.* 381, 154–167.

Washington, K. E., West, A. J., Kalderson-Asael, B., et al. (2020). Lithium isotope composition of modern and fossilized Cenozoic brachiopods. *Geology*, in press. doi: https://doi.org/10.1130/G47558.1

Wei, G.-Y., Wei, W., Wang, D., et al. (2020). Enhanced chemical weathering triggered an expansion of euxinic seawater in the aftermath of the Sturtian glaciation. *Earth Planet. Sci. Lett.* 539, 116244.

West, A. J., Galy, A., and Bickle, M. (2005). Tectonic and climatic controls on silicate weathering. *Earth Planet. Sci. Lett.* 235, 211–228.

Wimpenny, J., Colla, C. A., Yu, P., et al. (2015). Lithium isotope fractionation during uptake by gibbsite. *Geochim. Cosmochim. Acta* 168, 133–150.

Wimpenny, J., Gislason, S. R., James, R. H., et al. (2010). The behaviour of Li and Mg isotopes during primary phase dissolution and secondary mineral formation in basalt. *Geochim. Cosmochim. Acta* 74, 5259–5279.

Acknowledgements

This work and PPvS was funded by ERC Consolidator grant 682760 CONTROLPASTCO2. MD acknowledges support from a Natural Environment Research Council grant (NE/T001119/1), and AJW from National Science Foundation grant EAR-2021619. We thank Madeleine Bohlin and Xiao-Ming Liu for their constructive reviews.

Cambridge Elements ≡

Geochemical Tracers in Earth System Science

Timothy Lyons
University of California
Timothy Lyons is a Distinguished Professor of Biogeochemistry in the Department of Earth Sciences at the University of California, Riverside. He is an expert in the use of geochemical tracers for applications in astrobiology, geobiology and Earth history. Professor Lyons leads the 'Alternative Earths' team of the NASA Astrobiology Institute and the Alternative Earths Astrobiology Center at UC Riverside.

Alexandra Turchyn
University of Cambridge
Alexandra Turchyn is a University Reader in Biogeochemistry in the Department of Earth Sciences at the University of Cambridge. Her primary research interests are in isotope geochemistry and the application of geochemistry to interrogate modern and past environments.

Chris Reinhard
Georgia Institute of Technology
Chris Reinhard is an Assistant Professor in the Department of Earth and Atmospheric Sciences at the Georgia Institute of Technology. His research focuses on biogeochemistry and paleoclimatology, and he is an Institutional PI on the 'Alternative Earths' team of the NASA Astrobiology Institute.

About the Series

This innovative series provides authoritative, concise overviews of the many novel isotope and elemental systems that can be used as 'proxies' or 'geochemical tracers' to reconstruct past environments over thousands to millions to billions of years – from the evolving chemistry of the atmosphere and oceans to their cause-and-effect relationships with life.
Covering a wide variety of geochemical tracers, the series reviews each method in terms of the geochemical underpinnings, the promises and pitfalls, and the 'state-of-the-art' and future prospects, providing a dynamic reference resource for graduate students, researchers and scientists in geochemistry, astrobiology, paleontology, paleoceanography and paleoclimatology.
The short, timely, broadly accessible papers provide much-needed primers for a wide audience – highlighting the cutting-edge of both new and established proxies as applied to diverse questions about Earth system evolution over wide-ranging time scales.

Cambridge Elements ≡

Geochemical Tracers in Earth System Science

Printed in the United States
by Baker & Taylor Publisher Services